WHAT IS AI?

NANCY DICKMANN

BROWN BEAR BOOKS

Published by Brown Bear Books Ltd
4877 N. Circulo Bujia, Tucson, AZ 85718
USA
and
Studio G14, Regent Studios, 1 Thane Villas,
London N7 7PH, UK

ISBN 978-1-83572-047-9 (library binding)
ISBN 978-1-83572-053-0 (paperback)
ISBN 978-1-83572-059-2 (ebook)

Library of Congress Cataloging-in-Publication Data available on request

Text: Nancy Dickmann
Consultant: Matthew Lugg
Design and Illustrations: Square and Circus
Design Manager: Keith Davis
Children's Publisher: Anne O'Daly

Picture Credits
The photographs in this book are used by permission and through the courtesy of:
Cover: Freepik.com: **Interior:** Shutterstock: Mongkolchon Akesin 20–21, Mariangela Cruz 18–19, DC Studio 10–11, fizkes 16–17, Insta photos/Ground Picture 6–7, Monkey Business Images 8-9, nelen 14–15, Artem Oleshko 12–13, Drazen Zigic 4–5.
Artwork: Freepik.com.
All other artwork and photography
© Brown Bear Books.

Brown Bear Books has made every attempt to contact the copyright holder. If you have any information about omissions please contact: licensing@brownbearbooks.co.uk.

Websites
The website addresses in this book were valid at the time of going to press. However, it is possible that contents or addresses may change following publication of this book. No responsibility for any such changes can be accepted by the author or the publisher. Readers should be supervised when they access the Internet.

Words in **bold** appear in the Glossary on page 23.

Manufactured in the United States of America
CPSIA compliance information: Batch#AG/5666

Contents

INTELLIGENCE

What does it mean to be intelligent? There might be someone in your class who aces every test. Most people would say that this person is intelligent. And they probably are! But there is more to it. **Intelligence** is not just knowing a lot of facts.

UNDERSTANDING AND LEARNING

Intelligence is the ability to learn. It means understanding things. This includes things you can't see, like ideas. It's about being able to figure things out. It means seeing problems and solving them. Humans can do all this. We do it much better than animals can.

All humans are intelligent. People show it in different ways.

Kinds of Intelligence

Intelligence comes in many different forms. You need it to do all of these things.

Solve math problems and puzzles

Learn and speak another language

Understand other people's feelings

Sense rhythm and sound and create music

Know your own strengths and weaknesses

Picture objects in your mind

Think about answers to big questions, such as why we are here

CAN COMPUTERS THINK?

Computers and AI are not the same thing. Many home computers do not use AI.

Humans are really good at thinking. But what about computers? We use them every day. They're great for doing homework and looking things up. They help us send messages and buy things. They help us stay organized. But can they think like we do?

ARTIFICIAL INTELLIGENCE

People often talk about artificial intelligence. It's called AI for short. This is a new kind of technology. It lets computers think and learn like humans. AI can take in information. It uses it to make decisions and solve problems. AI is still pretty new. But it's improving all the time.

Computer Languages

Computers don't speak English or Spanish. They see the world in ones and zeroes. Scientists use special computer languages to create AI.

A **programmer** types instructions. They're written in a computer language.

The instructions get turned into machine code. This is long string of ones and zeroes. The computer can understand it.

The computer follows the instructions.

HOW COMPUTERS LEARN

How can you learn to ride a bike? An adult might show you how. They help you do it better. As you get older, you're able to learn on your own. You might read a book or watch a video. You know how to learn from your mistakes.

MACHINE LEARNING

Scientists have taught some computers to learn for themselves. This is called **machine learning**. They feed computers lots of **data**. They tell it what to look for. The computer does its best. It can remember its mistakes. It learns from them.

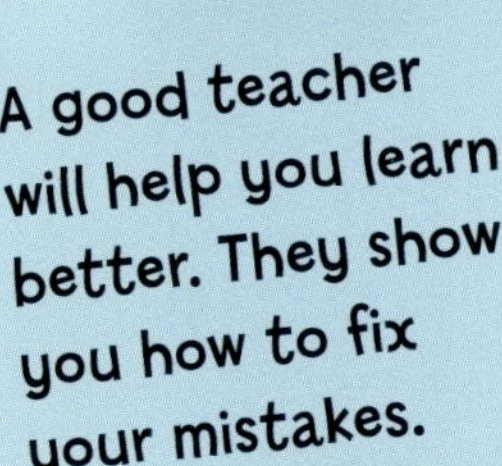

A good teacher will help you learn better. They show you how to fix your mistakes.

How It Works

Machine learning can teach computers to recognize things.

Scientists feed a computer photos of animals. Each one has a label, like "lion" or "snake."

The computer looks at all the lion photos. It figures out what they have in common.

The computer is given new photos. These don't have labels.

It **predicts** which one shows a lion. The scientist tells it if it was right.

The computer learns from this. Its next prediction will be better.

Following Orders

Many computers just do what they are told. They follow a program. This is a set of instructions to follow. But this is not machine learning.

WHAT AI LOOKS LIKE

You might already know what a human brain looks like. But what do you think AI looks like? In films about AI, there might be a shiny **robot**. Some robots use AI. But AI itself doesn't have a body. It's not something you can touch.

ONE BIG PROGRAM

An AI system is a really big, complicated computer **program**. It is stored on a powerful computer. AIs let you talk to them in different ways. Sometimes you type at the computer where it is stored. You might connect over the internet. Some AIs have a box that you type into. Then you get a text reply.

An AI system takes up a lot of memory. It's stored on big computers.

Showing What You Can't See

It's hard to show AI in a picture. Do an image search for "artificial intelligence." What do you get?

WORKING FAST

AI can "think" much faster than a human. This makes it really useful. It can go through huge amounts of data. It can solve complicated math problems. It can write reports and create pictures. For an AI, these tasks take only seconds.

SEEING PATTERNS

Humans are good at seeing **patterns**. Imagine that you see an unusual flower. Your brain compares it to flowers it's seen before. Even though it looks different, you still know it's a flower. AI can do this too. It does it very quickly. But it sometimes makes mistakes.

AI can quickly recognize faces. It checks many tiny measurements.

Helping Doctors

Some AIs spot signs of disease in scans of the body. They are faster than humans. Sometimes they find things that doctors missed. But an AI can also miss things. It might even spot the wrong thing.

Patterns Are Everywhere!

We can learn from patterns. Here are some ways that AI uses patterns.

Spotting and filtering junk emails

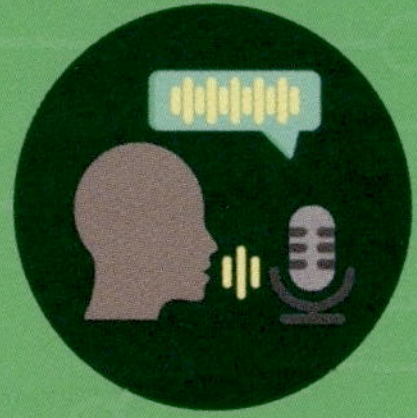

Recognizing spoken commands

Finding problems with factory machines

Turning handwriting into type

Suggesting films that you will like

USING AI

AI is already part of our lives. Many people use **chatbots**. These programs answer questions. You type what you want to know. Then the chatbot gives a reply. Chatbots can help write reports and emails. They can help with math problems. They can even write stories.

AI EVERYWHERE!

You might use AI without realizing it. Home smart speakers use AI. It helps them understand what you're asking. Social media **apps** use AI to decide what to show you. Online shops use it to suggest things you might like.

Finding Your Way

Map apps on phones use AI. It lets them show the best route. They help you avoid traffic or closed roads. They figure out how long it will take.

Fitness trackers keep track of your exercise and health. Some use AI to do this.

Helping Out

Do you use AI in your daily life? Here are some things it can help with.

- Improving the quality of the photos you take
- Keeping your home at just the right temperature
- Helping robot vacuum cleaners find their way
- Sending alerts to keep your home safe
- Making suggestions for music to stream

GETTING IT WRONG

AI doesn't always get things right. The answers it gives are sometimes wrong. And sometimes AIs just make things up! They are designed to be helpful. But they can't tell what's true and what's not.

TRAINED FOR SUCCESS

An AI must be trained well to work properly. This means giving it lots of good data. The facts must be accurate. There must be a good range of examples. Otherwise the AI will not give good answers. If it is fed a wrong fact, it might repeat it.

The Wrong Idea

Training data must treat all things accurately. If it doesn't, the AI will get the wrong idea.

An AI was given photos of many different vehicles. There were cars, bikes, buses, and trucks.

Only one photo showed a blue vehicle. It was a car.

The AI decides that all blue vehicles must be cars.

It is given a photo of a blue bus. It gets it wrong. It says that it's a car.

You need to think when you use AI. Don't assume that everything it tells you is right.

PASSING THE TEST

AI is designed to think like humans do. But could an AI pass as human? In the 1950s, someone suggested a test. A person would have two text chats. One was with a real person. The other was with a computer. Could they tell which was which?

IS IT ENOUGH?

If the person couldn't tell, then the computer passed the test. Today, many AI chatbots can do this. They give clever answers. These sound like what a human would say. But this only tests one kind of skill. It doesn't measure creativity. It doesn't test problem-solving.

Alan Turing made up this test. He also helped design the first electronic computer.

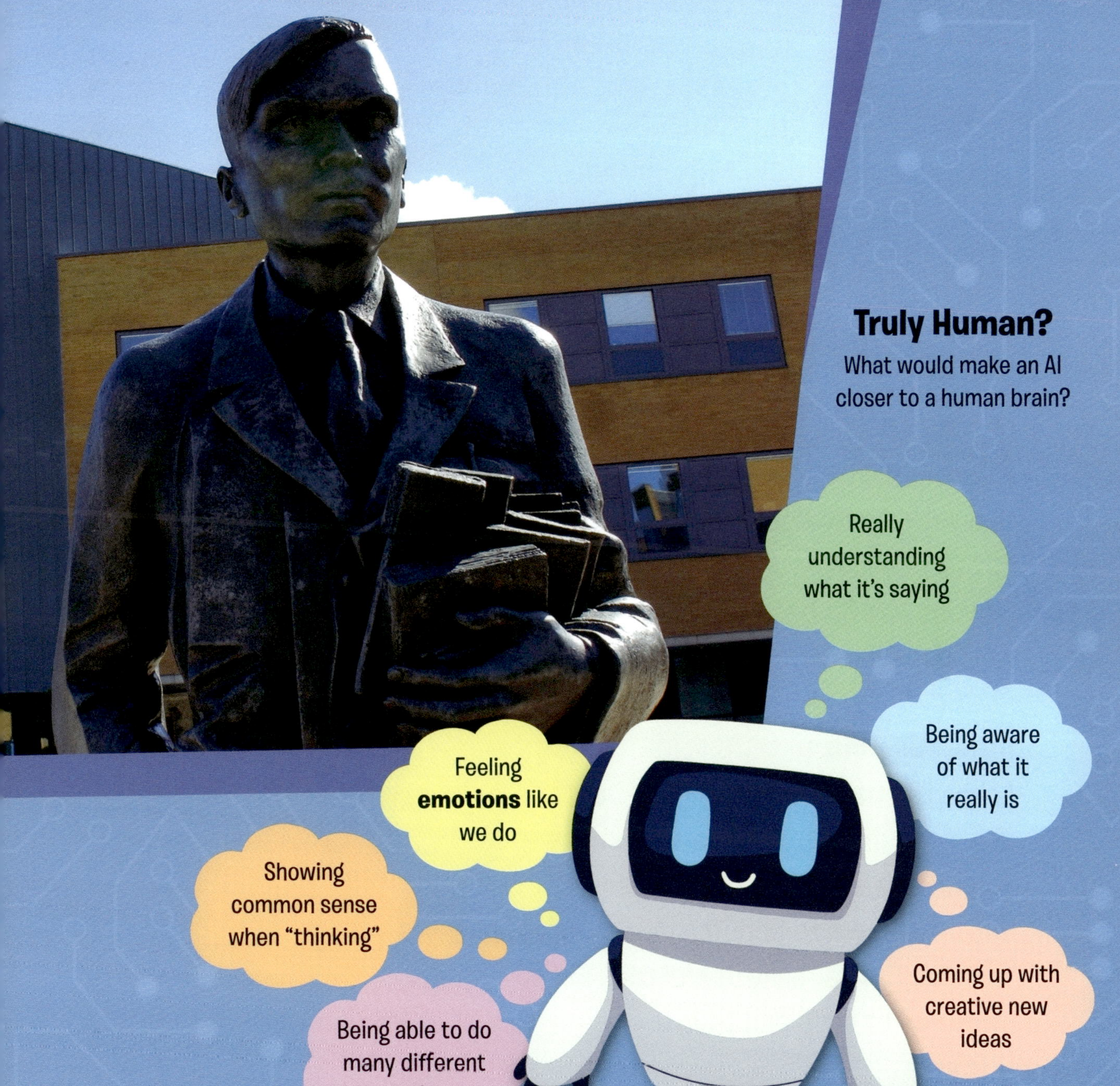

Truly Human?

What would make an AI closer to a human brain?

HOORAY FOR HUMANS!

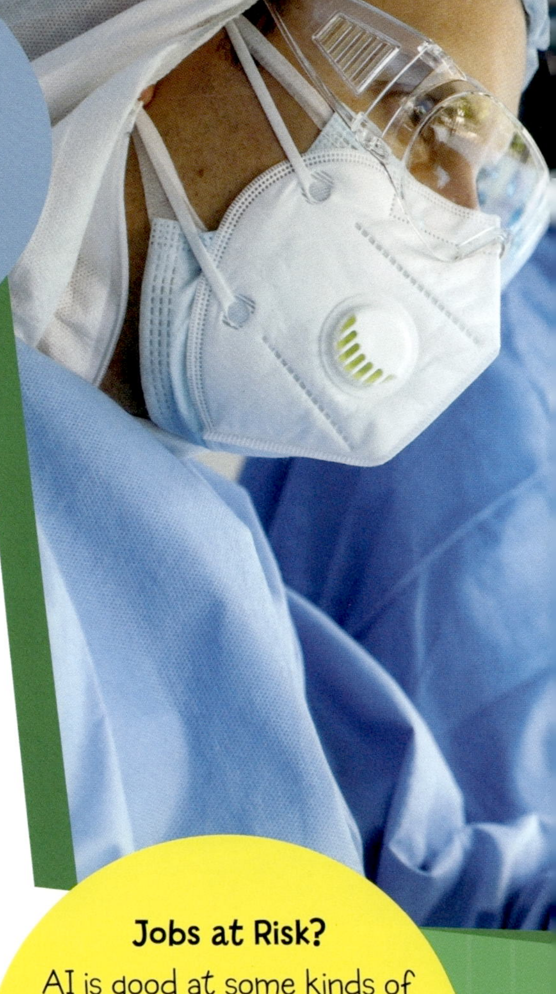

AI is part of our lives now. It does some jobs really well. There are other things that humans do better. AI is improving fast. But there may always be some things that humans can do better. Who knows what AI will be like in the future?

A USEFUL TOOL

AI has the potential to help us. Doctors use it to help diagnose patients. Scientists use it to find new planets. Business use it to become more efficient. This makes AI a really useful tool. But we still need to build and train it.

Jobs at Risk?

AI is good at some kinds of jobs. It does them fast and cheaply. Customer service help lines are one example. But if companies use AI to do them, people will lose their jobs.

Doctors often use AI. It helps them see how diseases spread.

The Human Touch

Some jobs need a human touch. They are probably safe from AI.

Teachers need to understand their students. So do coaches.

Therapists need **empathy**. AI does not have this.

Carpenters need to be good with their hands.

Artists, writers, and musicians must be creative.

Judges need a sense of right and wrong.

Inventors need to come up with new ideas.

QUIZ

How much have you learned about AI?
It's time to test your knowledge!

1. What is AI short for?
a. animal insight
b. applied industry
c. artificial intelligence

2. What do scientists feed computers to train them?
a. data
b. intelligence
c. pizza

3. Where is an AI system stored?
a. inside fancy robots
b. on big computers
c. in a human brain

4. Which of these jobs would be hard for an AI to do as well as a human?
a. solving math problems
b. putting films in categories
c. teaching someone to play the trumpet

The answers are on page 24.

GLOSSARY

app a computer program designed to do a particular job, usually on a smartphone or tablet

chatbot a computer program designed to have conversations with humans

data information that is stored or used in a computer, in the form of a series of ones and zeroes

emotion a feeling such as joy, anger, or fear

empathy the ability to know what someone else is feeling

intelligence the ability to learn or understand, especially in new situations

machine learning a method that lets a computer learn to do a task by analyzing lots of data

pattern an arrangement of similarities or trends between different items in a set

predict to make an educated guess about what will happen

program a set of coded instructions for a computer

programmer a person who uses computer languages to write programs for computers

robot a machine with moving parts that is programmed to do a job

text data in the form of written words

therapist a person who is trained to listen to a person and help them with their problems

FIND OUT MORE

Books

AI Basics. Elsie Olson, Lerner Publications, 2025.

Artificial Intelligence. Julie Murray, Abdo Books, 2021.

How AI Works. Lisa Idzikowski, Lerner Publications, 2025.

Websites

www.bbc.co.uk/newsround/49274918

kids.britannica.com/kids/article/artificial-intelligence/390648

softwareacademy.co.uk/ai-for-kids/

INDEX

Answers: 1. c; 2. a; 3. b; 4. c